The God Of Elijah

by

The Most Honored Elijah Muhammad
The Exalted Christ

Saviour's Day
February 26, 1959

Thursday Believers Meeting

Published by

T.H.E.M. Archive Society
(The Honorable Elijah Muhammad Archive Society)
t.h.e.m.archivesociety@gmail.com
(213) 973-THEM (8436)

Table of Contents

ACKNOWLEDGEMENTS

You are trying to make mockery of Me. But I ask you, how can a man retrieve the drowning man if he doesn't have a boat to go out in the water or a buoy to put around the man's neck? He has to leap in and battle the water and suffer as the man suffers. This is what I am doing. I am battling that which you battle. This proves the love I have for you.

We Forever, Thank You, Allah for intervening in our affairs here in North America. In the Person of Master Fard Muhammad to Whom All Holy Praise is forever due to you, for Your coming to redeem us who were lost. We Thank You for Raising up from our midst the Exalted Christ, The Messiah, The Most Honored Elijah Muhammad. We forever, Thank Both of You for Your Messenger, His Apostle, Our Comforter, The Paraclete, Your Reformer, Your Mercy, Your Great Redeemer to Us and to the whole of Humanity, The Honorable Minister Louis Farrakhan.

Thank You, Brother Minister Farrakhan. We would have never come to Know of Allah and His Christ had you not picked up their Mantle and Re-established his Mission to save us all. We Love you, Brother Minister Farrakhan. We Pray that Allah and His Christ, Blesses you, Keeps you, Restores you and Continues to Guide You and Your Family. We LOVE You, Mother Khadijah Farrakhan and we give Thanks to Allah for you and We Thank you for the sharing your Husband with us. We Thank Your children for sharing their Father with us and allowing us to have a Father Figure. We Love and Thank You.

Thank You, Brother Kenneth 3x formerly of Muhammad Mosque #7 NY, NY. For your entrusting me with your personal lessons and Archive of the Teachings of the Honorable Elijah Muhammad.

ACKNOWLEDGEMENTS

Thank you to my Loving Wife, Sister Jevedia Muhammad and Our divinely guided children Sister Mateenah, Sister Safiyyah, Sister Khalilah, and Sister Meme Muhammad.

Thank You, Sister Charisse Muhammad of Muhammad Mosque #27, Los Angeles, CA. for sharing your gift. Thank You, Sister Student Minister Aishah Muhammad, Brother Student Minister Charles Muhammad of Muhammad Mosque #32 Phoenix, AZ. and Brother Sabir Muhammad of Muhammad Mosque #27.

The Staff members of T.H.E.M. Archive Society: Brother Bilal Muhammad, Brother Jon Muhammad, Brother Michael J. Muhammad. Your Brother Eric Ahad Muhammad, I give Thanks to Allah, for you.

As Salaam Alaikum

INTRODUCTION

These are the transcribed words from the speech of the Most Honorable Elijah Muhammad that he gave on Saviour's Day. Thursday, February 26, 1959 in Chicago, Illinois

To the entire Family of the Most Honorable Elijah Muhammad. We Love you and send to you a heartfelt Thank you, for without you and the sacrifices of your family we would not be here today.

Brother Kenneth 3x of Muhammad Mosque #7, Who I had met in the early 1990's in Los Angeles, CA. at Muhammad Mosque #27. Brother Kenneth entrusted me his Light, as they would say prior to 1975. Now because of him blessing to me with these treasures I have been enabled to start on my journey on to the path of what I've come to believe is my purpose in this Life.

I pray that Almighty Allah and His Christ and Their Messenger in our midst are pleased with our endeavor. In the sharing of the Teachings of the Most Honored Elijah Muhammad with all of those who desire to read and hear His Invincible Truth from you Allah, Master Fard Muhammad to Whom All Holy Praise is Due forever. Through you, His Exalted Christ, The Most Honored Elijah Muhammad and you Their Messenger in our midst The Honorable Minister Louis Farrakhan.

We here at the T.H.E.M. Archive Society have chosen to do all that we can Humanely do to ensure that Allah, Master Fard Muhammad, that His Chosen One's from among us. The Most Honored Elijah Muhammad and His faithful Minister Louis Farrakhan. That their mirror destiny will be written in their words, shown by their works and with the Permission of

Allah, represented to the world through our works and our deeds, by our hands, as the believing Citizens of the Nation of Islam.

Master Fard Muhammad

"THE GOD OF ELIJAH"

In the Name of Allah, the Most Merciful, to Whom all Holy Praise is due, the Lord of the Worlds, the Most Merciful, the Finder and Life-Giver to us the Lost-Found mentally dead in the wilderness of North America.

My Beloved Brothers and Sisters, Ministers, Laborers I am more than honored to say to you that we are thankful to Almighty God Allah Who has allowed us to be here on this Great Day. Thanking you for your help and labor in this cause. And to the visiting Ministers, I don't know how many you are. Stand up. I want to say to you that I am happy and feel grateful to Allah for you and your great work that you are doing in the Name of Allah and for us and yourself in the Resurrection of our mental dead and I thank you.

My Beloved Brothers and Sisters who are visiting from the outside of Chicago, I want you to know that we, the Muslims of Number 2, Illinois are real happy to see your smiling faces among us and we hope that you will leave us smiling. I am real happy to see for the first time many of you that I have never seen before. And I am greatly honored by you and your coming to Chicago to visit us on this greatest day of our lives. And I am really small before these Ministers for the great things that they are saying of me. This is Saviour's Day, not Apostle's Day.

I feel real small to be sitting in the midst of you saying such great things about me. Why didn't you wait until I was somewhere else? That would have been better. I don't care about praise for myself. It is Him that sent me that you should praise. I thank you.

I want to ask again to see just how many that are visiting us from the outside. Every time I call a number, hold up your hand: Number 1, Michigan. Where is it? No one here. Don't have to mention Number 2. Number 3, Milwaukee, Wisconsin. Number 3 are you here? Not one yet. Number 4, Washington, D.C. Here we are from Washington too? Stand up you that are from Washington. Number 5, Cincinnati, Ohio. Are you here? None here from Ohio, Cincinnati at least. Are there any Muslims from Ohio, the State of Ohio? None. Are there any Muslims from Pennsylvania? Yes, here are two over here, two back there, thank you. New York? All Praise is Due to Allah. Stand up New York. Let's see what you look like. All Praise is Due to Allah. Lots of them from New York. You may be seated. I am so happy to see you. Are there any Muslims from Maryland? Hold up your hand or stand. Anyone from Baltimore, Virginia, North-South Carolina, Georgia, Alabama, Mississippi? I might call on Number 2 if I say Mississippi. Louisiana? Well, we have them from Texas and California. How about Missouri? Yes, here comes Missouri, thank you. And Kansas? None here from Kansas yet. We have Washington, State of Washington. None from Massachusetts? Stand up. Oh' yes. All Praise is Due to Allah. Thank you. And Rhode Island? None from Rhode Island. Connecticut? Yes. Thank you Brothers. We ought to have some here from Michigan. None from Ohio? Let's see about Idaho. Minnesota? All Praises is due to Allah. Come all the way from Minnesota here. We have a nice group of Muslims out there. Well I guess that is about all. I don't know where to begin, but I will begin again with you tomorrow if it be the Will of Allah and I expect to see you from then on. If it be the Will of Allah, I expect to.

My Beloved Brothers and Sisters, we are here today to learn Who is the God of Elijah? Who has spoken to Elijah? What does He look like? Where did He come from? Who is He? What is His Aim? Where is He now? Will we see Him? How old is He?

Where was He born? Or did He have a birth? We are here to learn of the God of Elijah. And Elijah is very happy to teach you of His God. We have lots of work to bring you into the knowledge of Elijah's God. It is not an easy job to teach you of the God of Elijah. You might think so, but it is a very hard job.

We call this day Our Saviour's Day; the birth of a Saviour for the so-called American Negro. You say, well why not for the world? We will answer that later on. We want to prove to you that you have been mistaken in what you have been believing or rather you were made blind, deaf and dumb to the Truth.

Today, February 26, 1959 is the Birthday of your Saviour and my Saviour. It is the Birthday of your God and my God. You will say, I don't want a God if He has to be born. We want to prove to you that you have been misled.

We are fast coming into the knowledge that there is nothing impossible for the man that is in flesh and blood to do. A long time ago when man had not advanced as far in the knowledge of science as he has today, it was impossible talk. You don't hear of such impossible talk today. We used to say that no one could go above the clouds, but God and or His angels, is that right? We actually believed that man could not fly above the clouds. We believed that he could not fly at all. And the only way that you could get above us was to go up on some high mountain. But today that old belief is now in the wastebasket. Now we see man ascending so far beyond the clouds until the clouds really look low, is that right?, Well, that was impossibility once.

Once upon a time the people did not think that they could go over the highway at a speed of 100 miles per hour in a horseless vehicle, but today is coming to pass that which we used to think or thought impossible. Now there is no more such thinking that we can not travel over the highways at the speed of 100 miles per hour or at speeds of 150 or 200 miles an hour without a mule or without a dog.

We are now moving around our planet so fast that we are now trying to think up something still faster than 100 miles an hour on the surface, so we want to get to our destination faster. We have now invented the plane. Now we fly 300 miles per hour, 1000 miles per hour. Alright, that was once impossible. Let us be careful and think over it. It is no more impossible.

It used to take us two or three months to go from America to Europe. No more does it take that long. It used to take six months sometimes; it depended on the good old wind. And in those days when the wind didn't blow, we were out there just motionless on the surface of the water, waiting and praying for the Lord to send us some wind to move the boat. But now, that is something in the past. We don't wait for no wind today. Today, the boat goes 30, 40 or 50 miles an hour and we used to sit around and just wait for the wind to take us. We couldn't imagine men traveling on the surface of the water at speeds of 35, 40, 50 and 60 miles per hour. Now we have little boats which go sailing over the water at 100 miles per hour for short distances. That was impossible once.

Well, now let us come hear about the God of Elijah. We want to know about this new God he has. What is this new Religion he has? Where did he get that kind of teaching? Who gave it to him? Who taught him that our Religion was Islam? Who has been teaching him that this is the Time of Resurrection of the dead? Why does he call us dead, when we have been taught that the dead are in the graveyard? We have been taught the history of 4,000 years ago. And we learned from this history that a man by the name of Moses, we will abbreviate his name, lived 4,000 years ago. Now what was that man 4,000 years ago to the Hebrew? A Saviour and an emancipator, is that right? That is what you call him, is that right? He was in every respect a Saviour for that people that you are reading their history of that he led them out from the house of bondage; out from under the hard taskmaster, Pharaoh, into another land wherein they could enjoy Freedom, Justice and Equality. A land of their own, is that right?

Mr. Moses was 4,000 years ago. Well, now you know who that man 4,000 years ago was. Alright, let's come back a little further. Now let's go beyond Moses. Now here God of that people before Moses come telling his people that the God of the Jews was named Jehovah, is that right? Well, if Moses' God was named Jehovah, and Jehovah said that He had not been known by that Name before; well, then let us question this thing a little bit.

If Moses preached to and taught his people 4,000 years ago that he had met with a God by the Name of Jehovah and none of his people had ever heard of any such God by that Name Jehovah.

According to the Bible, they had only known God by the Name Almighty, but by His Name Jehovah, He was not known to them, is that right? Alright, then how did the Hebrews accept a God Whose Name they had never heard of before from the mouth of Moses who was born in their midst? Now could they believe Moses? What proof did Moses show them that he actually met with the God Jehovah and that Jehovah was the God of the Fathers of his people and that Jehovah had come to deliver them?

He had heard their groaning and their moaning and had compassion or mercy upon that people and He had come down to deliver them, but He used Moses as His mouthpiece, is that right? Let us learn something about that. Let us learn of this God. Let us learn His Greatness. Let us learn the aims of the God Jehovah who was in the bush. Why didn't some of the Hebrews see Jehovah? If Moses met with Jehovah out there in the burning bush, what kind of Jehovah was He to be a God of Fire?

If I would tell you today that I met with God out there in a burning bush and didn't see nothing but fire, you, with your knowledge today would say to me, then God is Fire, is that right? Certainly you would. God is Fire because Elijah met in a Flame of Fire and heard Him talking out of a burning bush and you would declare from that day on that God is Fire. But the Hebrews never said that God was Fire, is that right? All Praise is Due to Allah.

We want to bring you up to Saviour's Day. Remember you can't do so unless we go back and get all that you really believe in. But we would like that you know the Truth of that One. If Jehovah says to Moses in a Flame of Fire, "Go to Pharaoh and tell him let my people go", what should Moses say? He should say, Listen, I have met Fire. These things you need to know. Today if I would say to you that I believe that God is Fire and you would say, No, not fire because we can take water and put fire out. Water is stronger than fire. And our God couldn't be fire because we master fire. Alright then. He would tell you, Well that was Moses' God. The little bush was all in flame and the bush did not burn. So the God had come down in the bush. That Flame of Fire was God. He must be Fire, but He just didn't want to burn the bush up. I am Jehovah, says Moses. "I am the God of their Father." Think over that. Abraham, Isaac, and Jacob. I am their God. And Jesus spoke of the same in His day and time that He is not a God of the dead, but of the living. Remember how He spoke to Moses saying that He was the God of Abraham, Isaac and Jacob? If I would come to you today with the same kind of talk, you wouldn't accept it.

If I were to say to you, I met God out there in a burning bush and prove that there is God. Did the Hebrew do that? They did not question Moses about who it was until he actually got them out of the land up in the mountains.

Then they wanted to know from Moses. Who are you going up in the mountain talking to? According to the history of Moses, he had a very hard time trying to convince them that actually he was sent from God and that God was with him and that he was with God. How much harder time should I have with you to convince you that I am of God and I have talked with Him, with your highly educated system of the enemy, with him to help keep you from believing? All of his wise scientific knowledge of physical things, yet you are not able to say with me in one voice, I believe as the Hebrews believed Moses that what you have brought must be from a God Superior to the God we have seen.

Let us make proof of this. It is not what the man just says from his mouth, but it is what the man can prove from his mouth. If we can't prove what we say, then we are wasting time. If Jehovah says to Moses, "Go to Pharaoh and tell him let my people go" and Jehovah was in the mountains to the people that He was freeing, think over that. Then if you are to be freed by God or by Jesus, as you think. You think today that you are to be freed by God or His Son Jesus, is that right? Perhaps you are already freed. Jesus freed you from the World 2,000 years ago. We will move up to that later on.

Now if Jehovah says: "Moses, I am the God of their Father. I love them. I hate Pharaoh. I hate his people because they have mistreated my people", but this people didn't know Him, is that right?

They never saw His face or form. They never heard His voice. They didn't know His Name. But yet He loved His people and His people didn't love Him because they didn't know Him. But they could bow down to the God of Pharaoh that Pharaoh had set up, is that right? They believed in all the God Pharaoh represented to them as you today. You have been and are today still being taught of a God that you can't see, is that right? A God that the slavemaster has been teaching you as Pharaoh taught the Hebrew. You know nothing of the God your own people were taught of.

Take time and let's be quiet and learn of the God of Elijah. If Jehovah had spoken to Moses and said, "Moses, I have heard my people moaning and groaning. I have seen wherein they have been afflicted. I have come down to deliver them." My friend what God will come to America? Will that same One come to America and say to you and I, "I have heard your groaning. I have heard your moaning. I have seen how that the white man of America has afflicted you for 400 long years, and I have chosen you, to deliver you from your hard taskmaster", would you believe it? Would you believe that was the same God that Moses was talking to? Would you believe that this was the same God that loved the Hebrews?

Well, if the same God loved the Hebrews so well that He got them out of the land of Egypt into a land of self-independence, a free land for themselves and in that free land, He sent to them His only Begotten Son today to pay for their sins against Him, well then, who will take you as a Beloved People?

Would the same God love the so-called American Negro who is 4,000 years from Him? And now He is befriending a people that did not even know Him and He parted seas for that people. He parted rivers for that people. He slaughtered nations for that people. He established that people in a land, so the book says, that flowed with milk and honey. He protected that people against their enemy, is that right? He fought their battles. He went before them and their armies and conquered nations for them, is that right? Well now, could He love the so-called American Negro good enough to conquer His enemy today and lead Him out of His struggle? Let's look into it.

Could you be loved as well by that God Jehovah? Think it over. He said that they were His people, is that right? And that He had come to make Himself known to them that they will be His people and He will be their God and He will never forsake them. If He is going to love you the same way, He must be going to forsake one or the other, is that right? If it took 2,000 years of this 4,000 years for that same God to give birth to a Son to save the people Whom he had forsaken, then this Jew had already been forsaken, although the God nourished them and babied them from Egypt to Palestine, is that right?

He babied them like a woman babies her young baby in the desert. He gave them bread and meat, and He sent bread from heaven to them because that is God that will create bread and meat for you in a place wherein there is no meat and bread. Would He then forsake you after you get in a land and then send a Son of His to die to make you live again? Think over it my friend. This is Saviour's Day. According to the Bible, He goes forth to make Himself known, is that right? To make Himself a Name. If the Jews failed Him, He didn't make Himself a Name. If the Jews forsake Him, He was not the God of the Jews and the Jews were not His people. And His Son came and denounced the Jews and said to the Jews: "I beheld your father fall from heaven like lightning. I know your father. Your father is the Devil." Think over that.

If His Son denounced the people that His Father had sent to reconcile to power. Alright, He brings about the God's love and the people's love for their God and yet He condemns that people as not being fit for no such love. And then He condemns His God for sending Him to die for a people that was created, even from the very beginning to disobey You. Because I know the history of their father. Their father You cursed him out of Heaven and You didn't send nobody after him.

I say my friend, you need to learn the Truth. It is not anything other than just the Truth that God has raised me up to teach you. There is nothing that I want to do but just to tell you the Truth. It is up to you to take it or Let it alone. But the Truth, you have never known the Truth. Why? Because of your modern Pharaoh and your modern Egyptians. This Pharaoh has ever kept the Truth out of your eyes that was good for you. With Mr. Moses 4,000 years ago, He said "I led them". He drove them out by the hand of His own Messenger, His Prophet; by the hand of a man. He brought them forth. Not with the hand of a spirit, but the hand of a man. All Praise is Due to Allah.

It is hard to tell you of a God that is in flesh and blood when Jesus teaches of a God, something that you can't see. You will believe in that kind of God. Why? Because Pharaoh said so. They said to Moses, "Who sent you? Who made you a Judge over us? Who made you a Judge?" They will respect any officer, any inspector of Pharaoh who sent proof. But now here comes a man speaking to them that Jehovah sent him. Who is this Jehovah? They didn't question Pharaoh. Think over it. Not once do you find in the history where they ever went to Pharaoh saying, "Who is the God you are teaching us to believe in?" Have you ever gone to Washington to question the President about the god that the Christian people taught you to believe? Why?

Because they have ever been your guide and that has been from the cradle. Therefore, the baby can't question the

Bible. He doesn't have the knowledge to do that. Mr. Moses, 4,000 years ago saved his people. That is what they say. Alright, he met with God in the burning bush. He met with his God in the mountains, is that right? He talked with his God in the desert. You believe all of that, don't you? Why? Because you read it in the book and the modern Pharaoh taught you that. And you didn't question him. You don't question the book because you never had no other guide, but just modern Pharaoh and his Bible, is that right?

But Elijah springs up in your midst to teach you of a God that you know nothing about. He brings you a Book from that God. And Moses brought a Book from Jehovah. Did Jehovah make the people believe that He was with Moses? How did He convince them that He was with Moses? He began to whip them, is that right? And the more He would whip them, the more they would realize that there was some God with Moses or behind Moses' work.

Regardless to the miracles, they saw right in their eyes; seeing water running down out of a rock, they did not yet believe that there was a God behind Moses, is that right? But when He suffered fiery serpents to come out and bite them, they yet didn't believe that there was a God behind Moses, that He was whipping them for their unbelief.

He laid scourges upon them many times, but they yet would not repent and say "Listen, I believe that the God of Moses is the God that we all should obey." They often time said, "Who provoked Him to anger? What are you doing?" Today you are tracing the same history. God is after you today. 4,000 years ago, Moses, Mr. Moses was on the scene. You believe without any proof. You don't see no manna falling from heaven today. You don't see swarms of birds flying up in your midst or going down in your frying pan. No. But yet you believe that happened 5,000 years ago.

I want to bring you to today. Mr. Moses, think over that. Who was the Prophet before Moses? No, they never had one before him. The Bible gives us no knowledge of a Prophet before Moses who led the Hebrews. Alright then, let's have some kind of reasoning.

If Moses was the first Prophet of those people and he prophesied that there had come to them the God of their Fathers and that he was with that God. He met that God. That God was backing him to lead the Hebrew out from under bondage and Pharaoh and that he would overcome Pharaoh and bring about a deliverance for his people. Moses did not have to go further. He just kept on the one subject: God has come to deliver you and you must get up and go.

What made them get up and go? What was the whipping that He sent to Pharaoh that made Pharaoh let His people go? Pharaoh paid no attention to the frogs and plagues and gnats coming up in his kingdom. He said these things always happened. He didn't pay any attention to the hail. He knew that he had hail before, is that right? He didn't pay any attention to that. He didn't pay any attention to the water being poisoned. In the season that would happen. He didn't pay any attention to the drought drying up his well; that could happen. But when death come to his house; he knew that death had been always, but this was a miraculous death.

All of these things I say took place 4,000 years ago. And you and I believed it, is that right? We want to move up to today. If you don't know who was the Prophet of Jehovah; you don't say that Jesus is the Son of Jehovah. You say He is the Son of God. I only want to get you in shape for Saviour's Day.

Let us see, now here is 2,000 years you know nothing about before Moses. That man's name starts with a "Y". Now you never have read that man's history. He led that same people for 2,000 years before Moses. And after Moses, we come to Jesus. We used the Name God Almighty for this Jehovah. 2,000 years later, we come to a man or Prophet, or Son of God by the Name of Jesus, is that right? We don't know too much about 2,000 years before Jesus, or 4,000 years, is that right? Because we don't know nothing about that, only what we read.

Now suppose the Hebrew didn't know anything about the Scripture. He had none, according to the Bible. They had never been students of any history of a Prophet before the coming of Moses. Alright, don't think that we are not living today in the same condition.

One Hundred Years ago, our fathers didn't have any schools. They were here under modern Pharaoh without any school. I would not teach this, breaking it down like this, but we have non-Muslims with us.

Well, if our fathers were serving the American Christian people for 300 years without any knowledge of the book called the Bible, without the knowledge of love. Then they have only been free for 100 years to study letters. Learn them and to read the Bible, is that right?

Well, now today, can they truthfully say that the Bible is God's gift to Moses or to Jesus? Can they truthfully say so? Or only by what their masters say? Alight so you say it is good for us. Then why did the masters prevent your fathers and my father from studying something that was good for 300 years? Have you ever thought over that? If our fathers were deprived of the knowledge of something that was written in the book that was good for the world of mankind.

Every knee should bow to the word of God in this Bible. Alright then, if it is so good now, what was wrong with it 300 years ago? These are facts I want you to look at as we go along. Now you boast that this is the word of God to me. Nobody beats you doing that. This is God's Word. No fool can tell me this is not God's Word. We are the greatest representors of the Bible. Fact about it, you are in front of your slavemaster. Less than 100 years up from slavery, you are now going ahead of your slave master representing his Bible and his religion directly from the mouth of God more than he is.

Since you are such strong believers that you are willing to go out and represent the white man's Bible and his religion as being directly from the mouth of God, what would you do if you actually had your real religion? And here you are representing that which you actually had not the truth of and have not had the knowledge of the truth, that it is truth. Would you do as Elijah?

Is that the beginning of the world, Elijah, 6,000 years ago? Is that the beginning of time? No. No sir. That is not the beginning of time, according to my God. And your God.

If Moses is as you say he was, the first Prophet, and now today modern scientists say that this old Earth is billions and billions of years old and that people were here before Adam, what have you then to say about it?

If we are to believe such stories that the whole Heaven and Earth were created just 6,000 years ago, My Beloved Brothers and Sisters, we are certainly disappointed. We must learn what it means when that is said. We should understand when we hear people talking about the world is 6,000 years old. Don't think of the Earth as being no older than that. Don't think of the Sun, Moon and Stars just 6,000 years old. The world that is 6,000 years old is the white man's world. Don't think of your people as being from Adam, that Adam is the father of all the people, and that he was just here 6,000 years ago. Adam could not have been your father and my father and the father of the white people. Then he could not have been our father because a white father can not get black children. If Adam was white, we are not his children. Let us move on over that because this may let us go back to trial and error and you won't be able to see into all of that.

Let us now leave this and come to Jesus. After Moses, 2,000 years later, God sent Jesus. What was Jesus to do? You say to save the world. What world? All of us?

If he came 2,000 years ago to save the whole world of mankind, then who is to save us today? And then if He came to save us 2,000 years ago, and has made God suffer for the world of mankind and dying and giving up His life to save the world of mankind; then why are we looking for Judgement? That man's death should have banished forever the thought of a coming Judgement, because the people are saved and don't need a Judgement to save people. No. You don't send innocent people to court to be judged. You send the guilty.

There will never be a Judgement, according to your own Bible, until the man of sin be revealed. The Jews don't need a Judgement because they are not guilty of sin. But the man of sin must be judged, according to his evil. All Praise is Due to Allah.

If Jesus did not come to save the world, what did He come for? If we remember that there are 6,000 years for this world to be divided into three; there we have three times to equals six, is that right? Well, if three times two equals six and there we have Moses 4,000 years ago, Jesus 2,000 years ago and another man was 2,000 years ahead of Moses, then the time of this world is up. If the time of this world is up, who is to be saved now?

If Jesus' death did not save the Jew and He denounced them and prayed to the Father to take good care of those who believed in Him; on His departure He prayed to the Father in these words:

"I pray not for the world, but for those whom Thou has given to me out of the world." Well now, what will become of the world? If he says to His Father, take care of those who believed in Him. Take them out of the world. What will become of the world? And who is the world? All Praise is Due to Allah. The Bible teaches us that in those days at that time when Jesus was born in Palestine 2,000 years ago of an angel carrying news that a baby was born and having the government upon its shoulders with peace and good will to all mankind. That never happened 2,000 years ago. If it happened, the angel had mixed instructions because it certainly has not come to pass. Think it over for yourself. Truth is Truth.

Again that old story, "He gave His only begotten Son to save the world." What fool-god would give His only Son to save the devil? This same Son rose up and denounced the very people that He was sent to as being nothing but the children of the devil and that His word had no place in them. Now today, you are still preaching that God gave His only Son that through Him we might live. The white race preaches that God gave His only begotten Son that they may regain their place in Paradise. Then what is that baby? Is that baby a Prophet? Is that baby the God? Is that baby to be sacrificed for the people? Oh, no. That baby is God Itself.

If the woman, while threatening with a baby, is seen by the Prophet to be clothed in darkness, that woman is not the Sun herself, but is clothed in the Sun. The Sun clothes the woman. Well, where is the power of our Day Sun coming from? Our day sun has a God over it, is that right? It is not the God itself. There is a God over the sun, because the sun was created, not of itself, but by the Will of God it was created. Well now, if the woman is clothed in the Sun, then there is something superior to the woman, is that right? And to the life of the woman. All Praise is Due to Allah.

Then if this woman is impregnated with a child and this child is to rule, then where is the rule of God if this child is going to rule? All Praise is Due to Allah.

We are just only questioning, bringing to you that which you don't understand and want to tell you that which has been hidden from your eyes and that which you were not able to question. I am questioning it for you and giving you the answer.

What is meant by the twelve stars? You couldn't make planets sit up on top of one woman's head. What is meant by the twelve stars? The twelve stars represent twelve men. There are twelve men, and these twelve men are twelve wise men. They are from the twelve tribes that make up the family called the human planet and they are the wisest of them. And why did the Prophet see her donned with these twelve men? It means that which the woman has. Their heads is with the head which she is pregnant with. That is what it means. They are the twelve major scientists, or twelve Imams. These are Gods.

These are like the One that you find here in the book that number 24 and that they found at the feet of Him that sitteth upon the throne and give up their authority. When a man wears a crown and puts it down at the feet of another man, he is giving up his authority and admitting to a Superior one for a successor, is that right? We only want to understand.

Know what you read there in the book. I mean this, that a woman caused wonder in Heaven because that type of woman was not born in Heaven. That kind of woman is not one of the heavenly family; but nevertheless, that woman is clothed with the sun and her head is crowned with twelve stars and is pregnant with a baby that is more valuable than all the other women's babies. This baby that she is going to deliver is a superior baby to the Angels of Heaven, is that right? That baby was not born 2,000 years ago in Palestine. It is not that Jesus that you find that they have in the book that was there in Palestine, it is another one. This one will rule. The woman was a woman that didn't belong to the family of Heaven.

That is why Prophecy says there is another one, a woman. The book don't give you inside secret of it because it was not time. It was secret. If this had been known to the world just 100 years ago, we would have issued the Judgment in about 50 years ago. But this is the Truth that was kept hidden. And the Truth don't like to be known at the time when the end is here and the time when it would be heard.

And at this time, everybody will again accept it; even the enemy of Truth will admit it is the Truth. Back in the older days, they had the old Bible that you used to look at. That old Bible, Family Bible, that the slavemaster first gave to you, that he used to pull out a long, long time ago and refer to it in that book. They used to teach out of that.

In the making of man, that one spoke in the spirit, saying to God, "If Thou make man, I will pay his debt", is that right? You have heard of that. God said, "If I make man, he will sin." And one of the spirits says; "If you make it, I will pay its debt 4,004, is that right? You just don't see this no more in your writing because the time was not in then. It is here now. 4,004. Let's see what. Now here we have 4,000 years ago. Then Moses was here, is that right? Alright then, if we have 4,000 years ago that Moses was here and that man is to pay the debt for that man of sin or that one that sinned 4,004 years, alright, we will just move on; put down the 4,004 years. Alright, from Moses, then today is the time to look for that man, is that right?

We have seen 2,000 years from Jesus, is that right? Jesus came 2,000 years after Moses. Alright then. Jesus did not pay the debt of sin because the world has been more sinful since the death or the birth of Jesus than it ever has been. So Jesus could not have been the remover of sin. Well then 4,004 years, according to the Bible, there would come a man or One to remove sin and to take a people out of sin that were justly guilty of sin and should by Justice be killed or chastised for sin. But this One clears them of sin by His coming and by His Justice, He clears a people of sin out of the world and throws that world of sin and saves that people out of that world who are equally guilty of sin, is that right? All Praise is Due to Allah.

If that man would appear today, which He has already appeared, 4,000 years after Moses; what people do you think answers that description today that He should take out of the world, cleanse them or forgive them of their sin that they did in that world; Take them out and then destroy that world that caused them to sin? What people do you know of today that answers that description? It only brings you to the door of Truth. If it is not the American so-called Negro, who is it?

The American so-called Negro was brought here from their native land and people 400 years ago. Alright, they have been following the white man and have been trying to imitate him in all walks of life, is that right? We are guilty, equally guilty with the white man in every sin that the white man is guilty of; also they are guilty of the same because they have been trying to practice everything the white man practices. They have no other guide but the white people. Then if God was to appear today and if He were to forgive that people who had been led by another people blindly, without knowledge of that people, without the knowledge of himself, and have now became a sinful people, should not it be the so-called American Negro? Were not we brought out of our own native land to a land that we did not know?

Our fathers were brought to a new nation in the Western Hemisphere. There they fall a victim to white man's slavery, is that right? There they were lynched and beaten and killed for 300 years without any mercy. And there they were taught the way of the white man and not the way of their own people. What did Ezekiel say about that? "And I will go after them, search for them wherever they have been scattered. I will bring them again into their own land. I will settle them in the midst of their own people and they shall no more go astray, nor shall they anymore be afraid. I will be their God and they shall be my people. Who answers this description other than the so-called American Negro?

Is it not true today that Africa even is much concerned over the American Negro? Over your plight here, they are all concerned. They are trying to bring themselves into their own independence and now are moaning over you. Nobody writes to you here from abroad telling you that you should be independent while you are yet a free slave. Seek independence for your people. What do you talk on for the future of your people? A job. That is what you were brought here for, to work. No leader of yours goes to Washington and tries to demand equal justice here for you in this part of the Western Hemisphere, is that right? Not one. He only asks for just that you may work for an honest living. That helps, true, as human beings, but that is not all, Brother. That doesn't make independence.

A man can be good to his slave and not give that slave any freedom for anything but just to do as he bids this slave to do and the slave will grow up children in his master's house and the child will seem as though he is at home. He can go in the master's house and feel he is one of the master's members of his family, is that right? And they think that the master should treat him as his own son because he was born there too. It is true that the so-called American Negro today says to America that we are citizens and we are human like yourself. You should treat us like yourself. You are asking the slave master to treat his slave as he treats his own free son and free daughter. That is what you are asking for. You are not asking for the right thing.

The right thing is this: Give me some earth or let me get where I can find some for myself. According to a statement made taken recently of our people in America last week in Jet Magazine, it says that our population here today is 19 million. Alright, 19 million so-called American educated Negroes not asking for a thing in the government of America, but just freedom to work on a job. All you are asking for is for him to treat me like you are treating yourself. Let me bear your sons and you marry my sons and my daughters and we will be brothers and sisters. That is what you want. You want to submerge your Black self in whites. What right do you have to go to the white man and tell him to treat you as you should treat yourself?

Why are you not going for self? You are too lazy to go for self and you are willing to sit down at the feet of the white man begging for bread rather than to go and make your own bread. You will serve your life long and your children's life long as free slaves rather than go for yourself, or even to go to him and demand a piece of land that we may stick up our little Ensign. Let us work there on that land, feed our own hungry stomachs, our own naked backs. Let us live together and keep our names as Black names. Let the world know that we think something of ourselves rather than running over telling the white man to clothe me so I can be white and blue-eyed like you.

You were not able to save yourself. The only thing you know was to follow them, the white race have not been nothing but a race of Devils. That you and I know.

I say, My Beloved Brothers and Sisters, A Saviour is born for you clothed in the Sun of Truth and with Freedom, Justice and Equality at His feet. He is going forth to do Justice. You are not guilty of the sins of the world. It is the white man that is guilty of the sins of the world. You are not the people who were born in sin and shaped in sin; you are not that people. You are the people whom that person has brought up and you are following that person as one of them, but you don't belong to them.

The white race was made in sin. They was made of sin, and that is all they have practiced since they have been on the planet Earth. It is sin and more sin, and you have become one here following them and you do as they do and now you hate right as they hate right. They hate a righteous person in their midst. They don't want him to preach Freedom, Justice and Equality. They don't want to hear it, hear the Truth. Why? The Truth will awaken you, open your eyes. They don't want it; so therefore, they hate you for doing such and now they are mad at you.

They hate a man of Truth. They will trail a man of Truth all over the country because they don't like for you to know the Truth and you will do the same because you don't know the Truth and you are only following your open enemy to do evil against the Truth without knowledge. So therefore, a Saviour must be born for you in Heaven. What was that woman? That woman don't belong to Heaven. She was not originally one of the Heavenly people. No. I will make One for them. He must be a made man to save you. You heard about it. You thought it was the man 2,000 years ago. If you make a body for me, I will go and die. You never did understand that. You have been confused over the teaching or meaning of that, my friend.

It means that before ever the man appeared in flesh and blood, there was a prediction of Him that He would come like that and that the Word of Truth is ever made manifest in flesh and blood. There the redemption will be made for a people. I will redeem them—read your book—I will redeem them and I will bring them again and pay the price. I will pay their debt. Think over that. I must have a body to go among them. Not that a spirit is coming falling out of space and needs a body. Their spirit is in the people. It doesn't have to come out of space.

The Spirit of Prophecy is the thing that now must be clothed with flesh and blood to bring about fulfillment of the spiritual prophecy. Do you understand it now? I hope that you do. Now, 4,004, the reality of this prophecy or this grant that is made that I will send One to them and He will save them. It is now made manifest in flesh and blood, but the flesh and blood that it is made in must be such that can go to that people and that people will receive it.

Therefore, you must make that Son part of one and part of the other. This woman that he saw in Heaven with the twelve stars around her head as a crown, it means those twelve wise scientists that agree with this One that we must have One to go after that one that is last. Let's make One to go after that one that is last and save that one from the evil world before we destroy it. How are we going to make that One? The Father says, "I know." So He goes and gets a woman out of that people, takes the woman up, puts her down and cleans the woman and made her fit to give birth to a perfect man. And that perfect man will look like the enemy and yet look like His people. Both people will receive Him because each people will see themselves in this One Man.

So He came not totally white, nor totally dark, but between the two, from both. Therefore, He could get among all of them. And He came to His own and His own know Him not. How could His own know Him when they didn't know themselves? This baby as He grew, according to Prophecy, His government will never come to naught. No King will ever destroy His government. He shall have no end. There will be no man as the guilty will no longer be going to war. There shall be no such thing as harm to the world.

The world has been in worse condition than it ever was. Alright then, that was not that Jesus 2,000 years ago; but Jesus coming at the end of 2,000 years after the death of that Jesus. That is the One. He is called Jesus by God Himself, by the Prophets and His Name, meaning the Emanuel that is God is with us, is that right? They have never called Jesus by that kind of name. He shall save His people from their sins, but not the Jesus 2,000 years ago. It is you today that God will save from your sins.

"I will forgive their sins and I will not mention them to you. I will just blot them out as though they had not been." He didn't ever say that to the Jews that He could do that. That Jesus said to the Jews: "Repent.", is that right? But this Man says to the people, "I will forgive your sins. I will wash you whiter than snow. Regardless to how much sin you have done, I will forgive all of that sin and I will declare you as Mine. You shall be Mine."

Shall God come to you as a man? In what way could He come to us that we would receive Him other than as a man? We will never receive anything like a God coming to us other than as a man. If He comes looking different from all other human beings, then He is not a human.

Well, we don't understand that thing. We don't even like human beings. You say, "God is not a Man." The Bible says God is not like a man. Who said that and what man is it that He is not like?

He is not the man of sin and is not like the man of sin. The Holy Qur'an says there is none like Him. It means this: There is none equal to Him. Never was His likeness, in Mercy, in Truth, in Justice and in Wisdom. There was not His likeness.

So Brothers and Sisters, don't reach out for nothing when you are reaching for something. When you make the God something that is nothing but just air or spirit without any form whatsoever, you are making a God for yourself that can do nothing for you. Nothing. A Saviour is born for the purpose of doing what? To redeem a people that is in that power, under a power of a people that they have given them-selves up to that people to be that people's servants. Now they belong to another people and therefore, the redemption price must be paid to get them from the people who do not own them; whom they have willingly subjected themselves to. Remember, for 100 years, the white man claims that he has freed you and I and now for 100 years, you willingly submitted to the white man to keep you.

You go up every year and vote for them to rule you, is that right? Every four years all of you go to the polls to vote for another of the white people whom you have agreed upon to vote for to rule over your kind. You are afraid to under-take any such thing as trying to set up a ruler for yourselves, right? You are afraid to do that. Certainly you are. Why? Because you fear the white man. Therefore, you vote for him to rule you. And never have one of them ruled you in Justice. No Justice has ever come to you from the rule of white people. I don't care which one you claim was good, he never was good enough to give equal justice and equality.

Abraham Lincoln would not have freed you if it had not been for his own purpose of bringing in a Union of the States. Otherwise, you would not have been free today. He didn't want to free you. He was brother and friend of his white brother, but he wanted one president over America. And by the white man of the south having millions of Negro slaves, he was able to oppose the north with these slaves. He would use them for an indefinite time. Fact about it, the north, if they had not liberated the slaves from the white man of the south, the south would have won the war. But the freeing of the slaves weakened the white south. My friend, remember that. Stop going around talking about Abraham Lincoln and that he freed us. Stop saying that, you are wrong. Look in the Pittsburgh Courier Paper. Read what J. A. Rodgers has to say about Abraham Lincoln and he gives you the proof.

So I tell you, my friend, you have been grievously misled. You have believed everything but the thing that was good for you. So today, God is with you. God is a Human Being. God is not a spook. Go see if the white people don't believe that God is a human being and that He has love for His people who are human beings. He can't because he is not one himself. You don't have love for cows and steers to stay in your home as members of your family, do you? Because they are not human beings, are they? You can't go out there and take them as one of yourself. They are not human beings.

When we fear a spirit somewhere, how can we get to this spirit? How can we get to be equal with this spirit when the spirit is not one of us? No. We don't want to go in heaven where there is nothing but spirit. We want to go to Heaven where there is flesh and blood like we are. We would have no joy in the spirit world and the spirit world will not have any joy in this world. What will a spirit world look like being dressed up in clothes or eating food? They can't do it.

So my friend remember, they went and searched for a woman that they took out of that race of people called the world of white mankind and cleaned that woman of her sin, used her, as it is said, cast out of Mary seven Devils to make her fit to give birth. Not a Jesus 2,000 years ago, but a Jesus for you today. Then that baby was brought up in heaven and taught by the Heavenly people.

The God of that Heaven was His, Who was His Father, also was a man. Then after He was of age, He was sent to the wicked world of the Caucasian people to study and learn their wisdom to conquer them and that man came about in 1877. He was born in the Holy City, Mecca on February 26, 1877.

The Bible condemns you in believing that Jesus was the One that was to save a people from sin because the Bible teaches us that that One will come from Heaven. Jesus didn't come from Heaven. Jesus came from Palestine. Have you ever thought of that? No, you are so drunk over misunderstanding that you never looked at that. No. I hope that is clear. There is so much more that I desire to take you through on this birth, but the time won't allow me and I am afraid you wouldn't be able to agree with me; but He is a man. He is in your midst and my midst.

What will He do when He comes? He will save His people from the sin of their slavemasters. From the sin of the Devil. He will destroy the world of sin and give the ruling power of the world over to those people that He shall choose. They shall rule the people forever under His Guidance.

He is Wiser than any God that ever lived. The Jehovah that you read of in the Bible, that the book is referring to that meets with us is not really the One Who met with Moses 4,000 years ago. He is not that same One of today. There was no such thing as One like that 4,000 years ago to deliver the people.

That Moses did the work of civilizing the white man while he was in the caves and hillsides of Europe. This man of Egypt is a symbolic Moses. Moses is symbolic.

There is a short history of Moses, so I was told, with a few hundred thousand of the Caucasian people who were in Egypt who were tricked by a King. In the Arabic Language, they called him Fir'awn. Fir'awn is the same one that they called Pharaoh. That man was once an Arabian and was so smart and tricky that he was about to trick the king there and the king caught up with him and the king gave him an opportunity to leave Arabia or suffer the consequences because he was so tricky. He was very smart. He asked to go to Egypt. At that time, it was not called Egypt, it was called another name. So he left Arabia and went to Egypt.

The man took some slaves, rather some Caucasian people and made them slaves in Egypt and kept them. These were not Abraham's seed. Remember, he took some white people, not out of the hills and cave sides of Europe, but these were some white people that had been left in Asia in Arabia when they were exiled into what we call Europe 6,000 years ago. About 4,000 years ago, the king got a hold to some of these white people. They were serving our people in Arabia. There the Arabs kept them ever since they were created on Pelan by their father Yacub, and when the king gave orders to drive out all of the Caucasians into West Asia as they now call it Europe. These Arabs hid them in their homes, some of the white people that they thought would be alright and they would keep them there and let them serve them as servants.

These are the people that Pharaoh got a hold to. He got a hold to some of these and when they got into Egypt, by then he made slaves out of them. There was only a few hundred thousand, not a whole nation and that is the few that were made an example for our own enslavement here; or rather they used it as an example for that which is to come. Actually, the race of the Caucasian people were in Europe.

At that time, they were walking around on their all fours. At the time of the birth of Moses, they were living in caves and up in trees and what not. They were wild people. They looked just like a monkey that you see in the zoo. That is the way the white race looked in the time of Moses in Europe. They had no knowledge of themselves. They had been there for 2,000 years, exiled by the Muslim people of the East. They cast them out like the Bible says. God cast them out of the Garden known as Arabia at the time and put them in a place where they had to work for their living in thorns and in thistles.

That was in Europe they drove the Adamic people, driving them out of the Garden of Paradise. The white race was driven out. They started trouble there among the Holy People, making the Holy People to make war against each other; and finally the old red dragon in Heaven made war and he was cast out. Where you read that, that was the white people that was cast out of Asia 6,000 years ago.

The same thing must come about here today. They must be cast out today. There is no such thing as the white man having peace in Asia, is that right? No such thing. And therefore they will have to be separated from Asiatics in order to bring about peace. You had to have a Saviour born. You have to have a God with you. That Saviour or that Son is God Himself. He shall rule at the end of the time of the white race. Then this Man comes to you and I.

The Arabs call Him Allah, meaning that He is God in Person, and they also give Him other Names. He has hundreds of names in their language. Your Bible says that His Name no man knows but He Himself. Alright now, if no one knows Him but He Him-self, let us take good notice of that. If a man comes to us and no one knows His name, then God was not known, is that right? No one knew the real name of God.

Today I preach to you that God is Allah. Alright, Moses preached to the people according to the Bible, that His Name was Jehovah. Let us remember these things. Let us not argue with each other. His Name was Jesus; His Name was so and so. If Jesus was born in Palestine 2,000 years ago, as we have learned here, He did not save the world from its sin. Well then, we must stop looking back at that man's history as the Saviour of the world, is that right?

We say that He will come again. Where is our proof, when He prophesied the coming of another one? If He said that His Father will send One, well then alright, He has declared to us that His Father sent Him and declared to us that He has no power to do anything except it was from His Father, is that right? Well then, what right have you and I to be sitting around here talking about we are waiting for His return? We are foolish, miserable, ignorant children of God. We don't understand the scripture, that is the trouble. He is not coming back. He is gone like all the other Prophets and the last one to the Jews, but this One that He prophesied that would come is God Himself, that would come, and the One that would be sent is from that God.

In the Judgment, He will have One with Him and His work will be among the people and it will last between 20 and 70 years. From 20 to 70 years, His work will go on and He will have a disciple, or One that He calls His Messenger. You have this in the last book of the New Testament.

Behold, I will send My Messenger before My Face. Not that I will reveal a message to One. If He is going to stay in hiding, He would say, I will reveal My word to such and such One and He will say to you, so and so. But when He says, I will send One, I will be in Person and I am going to prepare that man Myself. I will send Him from My Own Self. I will put his words in his mouth and he shall say to the people all the Lord shall say unto him and without the word of God, he will say nothing.

That is a man totally sent directly from a Man and given orders directly from that man to carry out those orders as ordered to carry out these orders. Now this man is saying only what he is going to be told. He doesn't say He revealed it; He has shown me in a dream, He has shown to me this vision. He doesn't say these things. He says, the Lord says so and so, and you shall do so and so and so. He speaks more forcibly and with more authority than a man who has his revelations through visions.

And I want to say to you that God has come to you and I and He was born in Heaven. He was not born in the white man's world. He was born in a Holy place where white people are forbidden. They don't allow Christians in the Holy City Mecca. And yet Mecca is the most Divinely defended city. You can't make war against Mecca. You can't attack it and win. The white man will tell you that. You can't fly over that city, and you can't take that city. That is a city that just can't be taken. It is a city that is defended, by Almighty God Himself.

And right there lives their righteous. Their righteous come here. When you hear tell of these people, they are good people and they are going there to increase their goodness. The white man's books teach that it is there in that land in a desert-like country. There for the first time you come face to face with your God, for the first time. No white man will deny that. No white man will say that I am not telling the truth that white people can't go to Mecca.

Even American planes helped carry some of the pilgrims near to Mecca, but they are forbidden. It is the Capital of the Muslim World. It is our Capital, and we forbid you if you are evil to come in our city. It is a city that is defended by a power that is superior to the white man's power. It is defended by the forces of nature.

There once the Christians of Abyssinia tried to attack Mecca with an army of elephants and he lost. The Jews and many times the white man has tried to make an attempt to get in and some of them will tell you if you happen to disguise yourself and try to get through by some divine means, he says you will not be able to cross the threshold. You will be torn to pieces.

One Muslim told me one tried to disguise himself and he got in among the throngs and dogs were made aware of his presence and he said they come up to this man while he was making a pilgrimage to Mecca himself; he said that these dogs came through the crowds just walking around smelling people's legs. And after while they got to this devil that was disguised trying to get by and both dogs jumped on him and started eating him up and when they captured the dogs and stopped them, they said they learned that he was a Caucasian. They had to take him away. Well, they killed him anyway.

This is the city where the Bible teaches you that one woman was sent. That one woman was sent into Mecca. That is the place where she gave birth to your Saviour and my Saviour now about 81 years ago. I think it is 1877, this is 1959. He is 82 years old today and He will live a long, long time. He will be 400 years old and the same as He is today. I can prove to you by your Bible. His age is there in the Bible.

I was with Him for around three years. He taught me night and day and He gave me to you as His Father gave Him to us. He gave me to you for the purpose of bringing you to Him. That is my job, to bring you to Him because you and I belong to Him.

The three years that you read of in the Bible that Jesus suffered; it is this One that suffered three years and one-half in America for you and I. He suffered three years and one-half under the same treatment you and I would receive. He didn't try to protect Himself from any ill treatment of the white man, given to Him out of justice. He was prosecuted, put in jail just like you and I before ever we were brothers and the white man ordered him out of America, but you don't see nothing in the papers. I was right by Him when He was down. I was with Him day and night. I went to the prison house in Detroit and in this city too, after Him. He would send for me just for me to learn what I had to go through. He was well able to defend Himself, but He wanted to show me what Myself and My Followers would have to go through, and that He loved us so well that He would suffer that first and He did so.

The white man knows Him, but they don't want to talk to you about Him. He has now secreted Himself in certain places and He is taking care of them for prosecuting both of us. They know all of this. The white man is well aware.

That is why they are so concerned over you coming to this kind of meeting. The white people don't want you to visit this kind of meeting. They don't want you to visit Muhammad at all. They rather that you go and become any other follower than to become Elijah's Follower because they have learned that Elijah is teaching you the Truth.

If you notice you will see the F. B. I. driving around watching us as though we were bank robbers. They are watching to see who is going to accept the Truth and who is really among us. He wish that he could put a stop. These are Federal agents. These agents are the ones who are trying to frighten you away from the Truth, and they are successful in doing it. There are thousands of our people right now who dare not come to this meeting because they fear the F. B. I. The F. B. I. sent me and my followers in '43 to jail for 5 years, but that didn't stop me. They didn't stop my followers, they are still here.

All Praise is Due to Him Who was born on this day to bring life to us sitting in darkness. It is by His Light of Truth that we were able to go and serve those three to five years and come back home with a smile. We continue this and we have been denying that it is a burden to us. We don't feel burdened from no such penitentiary. We feel happy that we serve that time. I don't know that I served no such time only by just remembering. I have no effect from it. And I made more converts in jail and Federal penitentiary than I did on the outside.

If you will be at 4100 South Parkway at the Metropolitan Community Church tomorrow at 2 pm, Saturday, and Sunday, you will learn much more of what I am saying to you. You will see many of our people there tomorrow, Saturday and Sunday. If you can't make tomorrow, don't miss one day. Tomorrow we want to begin telling of our Temple progress. There will be reports from various Ministers of their followers and their

Temples. They will be there reporting tomorrow, and you should come and hear what they have to say.

Also you will learn how the prison houses are accepting Islam over Christianity. I sent to them this week, yesterday, a gift of $440 for this date. I gave those 88 prisoners $5 money orders a-piece. 88 of them. It is not the first time. Year before last I did the same thing. Now we do that because of what? He that was born on this Day, our Saviour, has put love in our hearts for our brothers. The first time that I ever knew love for my brother was when He that was born on this Day opened my eyes to the Truth. Now today I love my brother and I don't mind living and dying for my Brother Black Man.

I thank Allah for sending us a Saviour. And He said to me after 3 ½ years' suffering among us, He said, "Your debt is now paid. You shall never desire that which you cannot reach anymore." And I thank Him that His words are Living Truth. He said to me this: He said, "I give you the Key. I will not go after anyone that you turn out." I began to think deeply when He said that. I shall be careful who I turn out. He says tell the preachers if they will submit and allow you to teach them, they can have their churches and their followers. Tell them you don't want to take their followers away from them, but to allow you to teach them the Truth so that they can teach their followers the Truth. Tell them that you will give them more money, more good homes than they ever had; and He said to me, Tell all that will believe and follow you and believe and follow Me, I will set them in Heaven at once." He said "money, good homes, friendship in all walks of life," and that stands true today.

It is forever being made manifest among you and I that believe. Remember, my Beloved Brothers and Sisters, that you are no more a forsaken people. You are no more at the mercy of the white man to be treated as a dog. If you come on here to

Islam, my Beloved Brother and Sister, you will find power against that evil people who are against you. God will give you power against them. But you are at their mercy as long as you try to do as they do and believe as they believe. You are at their mercy. But Islam, the God of Islam will grant you power against them. They will never be able to do to you what they used to do. Fact about it, they will be afraid because they fear Islam.

They fear that and they fear Him. Did you not know that America and practically all the white world don't want any words with the Islamic people?

They will tell you that once Islam and Christianity meet on the battlefield for the showdown, that will be the end of the Cross. Look at the many millions. Look at our people going away from white people. Do not you see how they are giving up white man's rule and Christian religion and going for themselves? All Africa today is in an uproar over the power of the white man. There are many throughout Africa and Asia just about got rid of the white man. The white man is no more in India, no more in Pakistan. He no more rules over China. Well now, what are you and I sitting around here waiting for? We must remember, Brothers and Sisters, that this is the end of the white man's world of power. And God is on your side and on my side to take our place in the Sun and rule as people should rule a people in Peace and Righteousness.

I thank you for your patience here this afternoon. And remember that this Day 82 years ago, the 26th of February 1877, the Jesus that you have been thinking was born in Palestine for you and me was born in Mecca, and today is seeking your deliverance and will win. He will deliver you from this people here. I am so happy and thankful to see your faces here this afternoon. We have the whole weekend of nothing but meetings every day. Be sure to be at the Metropolitan Community

Church 4100 South Parkway tomorrow afternoon and Saturday at 2 P. M. Sunday will be our last day here. That is what we want to do.

Fact about it, from tomorrow on, we have to do for the so-called American Negro. We want to put you on top. We want you to get together. You can never have any future with others as an independent people under Christian rule. So we want you to be out there to hear and learn Who was born to save you and I that believe from the power of this world and from its destruction. The white race today is expecting total war between themselves and Russia at any time. They expect total war, and even war that will end war.

I say to you my Beloved Brothers and Sisters from my heart, Join on to your own people in Islam. Join on to Allah and Islam and save yourselves from the destruction of this world. You don't have to pray for forgiveness of the sins that you committed before coming into the knowledge of yourselves. Up until you Learn of Islam, all sins that you have committed on the Planet Earth are forgiven. Come out to the Temple of Islam. Come here and Learn of yourself. And let no one frighten you away from your own salvation, regardless to what people say. As long as you see us here, come on.

So at this time, I say to you, my Beloved Followers and visiting Brothers and Sisters from the City of Chicago, I am thankful to meet you out here this afternoon and I hope that you have learned something from what you heard. We hope to get better acquainted with each other.

Every day we seek your acquaintance, and if you will come among us, you will learn that we are not a people that you were taught that we were. We will teach you how to live and Lead the best and most peaceful happy life you ever lived. We teach you to keep away from spending money on medicine and doctors, as you have been in the past; teach you to save your money by refraining from throwing your money away in all of this no-good dope and what not, bad food and drink. And you waste a fortune in poison.

Unite at this time. Unite together with us and the old Babylon, the old world will fall and you will be Left in the clear. I thank you my Beloved Brothers and Sisters. I have quite a few of you that I am inviting for dinner with my wife and my family and so I am going to say to you that I won't be able to feed every one that is visiting with us. But each day, I will have some of you at my home because I would like to meet you personally and I would like to listen to you tell me something. Don't sit down to listen to me because you listen to me and hear me every day. So when you come to my house, I am going to listen to you. I want you to talk to me and tell me what you know.

Thank you very much and may Allah's blessing and peace be upon you. And all that are hearing this Truth, may He bless you forever, as I say unto you in the Name of Allah, As-Salaam Alaikum.

The Most Honored Elijah Muhammad
The Exalted Christ

JULY 30, 1972 (excerpt)
Theology of Time Lecture Series

"I'm looking for help; I'm looking for the black scientist that knows something about how to build and how to tear down. I want both a wrecker and a builder. You will be your own ruler. I will teach you how. As you see, I'm my own ruler. And those that follow me they are coming along fine. They soon can go for themselves. Some of them are already going for themselves. This is what I want to do: I want you to be yourself and rule yourself. Don't jump out before you get your diploma, lest you come running back to me. Wait till I tell you that you can go for self. Don't try before you get the diploma or your degree. If you have already gotten a diploma from some part of this work or on some part of this work, now get your degree."

"I want you to remember; today I have one of my greatest teachers here...(sycamore tree talk)" "We have with us today our great national preacher. The preacher who don't mind going into Harlem, New York, one of the most worst towns in our nation or cities. It is our brother in Detroit or Chicago or New York. But I want you to remember, every week he's on the air helping Me reach those people that I can't get out of my house and go reach like he. I want you to pay good attention to his preaching.

His preaching is a bearing of witness to Me and what God have given to me. This is one of the strongest National Preacher that I have I the bounds of North America.

Everywhere you hear him, listen to him. Everywhere you see him, look at him. Everywhere he advise you to go, go. Everywhere He advise you to stay from, stay from.

So, we are thankful to Allah for this Great Helper of Mine, Minister Farrakhan. He's not a proud man. He's a very humble man.

If He can carry you across lake without dropping you in, he don't say when He get on the other side, you see what I have done? He tells you, to See what Allah has done. He don't take it upon himself. He's a mighty fine preacher. We heard him every week and I SAY, CONTINUE TO HEAR OUR MINISTER FARRAKHAN!"

AL-NAJM "The Star"

(Revealed before Hijrah)

With the name of Allâh, the Most Gracious, the Ever Merciful (I commence to read Sûrah Al-Najm).

1. I call to witness every part of the Qur'ân when it is revealed,

2. That your comrade (Muhammad) has neither deviated (from true guidance (in his practices) nor has he erred (in his beliefs).

3. He does not say anything out of (his own) fanciful desire.

4. This (Qur'ân) is nothing but (pure) revelation, revealed (to him by God).

5. The Lord of Mighty Powers has taught him

6. (The Lord) Whose powers become manifest in manifold and frequent ways, with the result that he (- this Messenger of God) attained perfection and fullest vigour (both intellectually and physically).

7. And he attained to the zenith of heights (in his spiritual ascension).

8. Then he drew near to Him and afterwards he descended (to humankind, for their guidance).

9. So that he became (as it were) one chord to two bows or closer still.

10. Then He revealed that excellent and mighty (Qur'ânic) revelation which He had to send to His servant (Muhammad),

11. Whose mind made no mistake in (the interpretation of) that which he saw (during the ascension).

12. Will you doubt and dispute with him concerning that (sight) which he saw with his own eyes (it being no figment of imagination)?

13. And, of course, he saw Him (in His another manifestation to him) yet another time.

14. It was near the Sidrah, which stands at the farthest end (of knowledge),

15. Near where also is the Garden which is the real eternal abode.

16. (This was) when the sublime thing (-the Divine Manifestation) which was to cover Sidrah had covered it.

17. (When he saw the Divine Manifestation) his eye deviated not (from the certainty of the Truth) nor did it wander away (from the invincible faith on which he stood).

18. (It was the moment when) he saw the greatly important signs of his Lord.

19. Have you had a look at Lât and 'Uzzâ (the gods of idolaters),

20. And another, the third (goddess of no account) Manât?

21. What (an ignoble idea you have)! Are you to have the sons and He the daughters?

22. That indeed is an unjust division (and unbecoming from your own point of view, which looks upon the birth of a daughter with condemnation).

23. (The fact is that) these are mere names (bearing no significance) which you have coined, you and your forefathers, for which Allâh has revealed no authority (for their being worshipped as gods and goddesses). These (idol worshippers) follow nothing but mere conjectures and the fancies of (their own) minds. (They do this) even though true guidance has already come to them from their Lord.

24. Can a person always have whatever he desires?

25. Rather (all the blessings of) the Hereafter as well as (those of) this present life belong to Allâh.

26. Behold! So many an angel is there in the heavens but their intercession can be of no avail except after Allâh has given (them) permission regarding him whom He wills and with whom He is pleased.

27. (Verily, it is only) those who do not believe in the Hereafter and (who) give feminine names to the angels (because they ascribe them to Allâh as His daughters);

28. But they have no knowledge of the matter. They follow nothing but conjecture, yet by no means can a conjecture be a substitute for the established Truth.

29. So turn aside from him who turns away from Our remembrance, and seeks nothing but the present life.

30. That is the highest attainment of their knowledge. Verily, your Lord knows very well those who go astray from His path, as He knows very well those who follow the right guidance.

31. And all that lies in the heavens and all that lies in the earth belongs to Allâh alone. The result of this is that He recompenses

the evil doers according to (the evil of) their deeds and rewards those who do good deeds with the fairest reward.

32. Those who avoid grave sins and open shameful deeds but (who are sometimes guilty) only (of) minor offences (will find) your Lord is Lord of immense protection (and resorts to His mercy in such cases). He knows you full well (since) when He created you from the earth and when you were embryos in the wombs of your mothers. So make no pretensions to the purity of your souls. It is He Who knows best who (truly and fully) guards against evil.

33. Have you taken notice of the one who turns away (from guidance),

34. And gives but a little (in the cause of Allâh) and does it grudgingly.

35. Has he the knowledge of the unseen so that he can see (his own future)?

36. Has he not been informed of the contents of the Scriptures of Moses,

37. And (those of) Abraham who thoroughly and faithfully fulfilled (the commandments of his Lord).

38. (The Scriptures say that) no soul that bears a burden shall bear the burden of another (soul).

39. And that a human being will have (to his account) what he strives for.

40. And that his strivings shall necessarily be seen (and evaluated),

41. Then will he be recompensed fully and fairly.

42. And that to your Lord is the eventual return.

43. And that it is He alone Who makes people laugh and weep.

44. And that it is He alone Who causes death and brings to life.

45. And that He Himself created the two species - male and female;

46. From a sperm-drop when it is emitted.

47. That it is He upon whom rests the raising (of the dead) again to life.

48. And that it is He Who fulfills the needs and endows (the gift of) capital contentment, and grants wealth.

49. And that it is He Who is the Lord of Sirius (- a star worshipped by some polytheists).

50. And that He destroyed the ancient (tribe of) 'Âd;

51. And (the tribe of) Thamûd, so that He spared not a single soul of them.

52. And (He destroyed) the people of Noah before (them). They were extremely wicked and most rebellious.

53. And (that it is He) Who totally pulled down the subverted cities (- Sodom and Gomorrah of the people of Lot),

54. So they were completely covered with that which overwhelms (when the punishment befalls).

55. (O human being!) Which of the bounties of your Lord will you then doubt and dispute?

56. This (Prophet) is a Warner (towards all the peoples of the worlds) from among the (series of) Warners (to one particular people) of old.

57. (The Hour of punishment) that was (promised) to come has drawn nigh.

58. None can avert this (doom) besides Allâh.

59. Do you then wonder at this announcement (and yet pay no heed to it),

60. (For on hearing the mention of the Hour of punishment) you laugh rather than weep.

61. And you remain proudly heedless and haughty.

62. You should better prostrate yourselves before Allâh and worship (Him).

The
Holy
Qur'ân

EXPLAINED BY
NOORUDDÎN

THE ORIGIN OF BLACKNESS - STUDY GUIDE 21

By The Honorable Louis Farrakhan

In The Name of Allah, The Beneficent, The Merciful. We give Him praise and thanks for His Mercy, His Goodness, His Guidance in the form of Divine Revelation that comes to us from the mouth of His Prophets and Messengers. We thank Him for raising up among us One to lead, teach and guide us to the straight path of God, the Most Honorable Elijah Muhammad.

I greet all of you, my dear brothers and sisters, with the greeting words of peace: As-Salaam Alaikum.

It takes great strength to abide in Wisdom. We are a people used to foolishness and we do not enjoy being challenged to correct behavior that God is displeased with. As long as the lecture allows us to enjoy something that tickles the ear—but does not raise the level of consciousness—sometimes those kind of lectures please us. But when you want to make people responsive to The Call of God; to be better than we are and accept a responsibility that will make us great men and women, you find that many are called, but just a few are chosen. I hope you will not be like that, because Allah (God) is completely Independent of us.

It is good for us to learn and grow and to have clear criteria by which we can measure ourselves and measure others. We are talking about The Coming of God to a specific people to accomplish a specific purpose. These people, though living, are considered dead. The Bible and the Holy Qur'an are not referring to people in a cemetery being raised by the blowing of a trumpet. This actually refers to human beings that are spiritually dead in the grave of ignorance, and hear a Truth that is like the sound of a trumpet. When a trumpet is blown, even if you want to sleep, it becomes difficult. The trumpet is a brass instrument and the sound of Truth in the ear of one spiritually dead would be like the brassy sound of a trumpet. But there will be a warning in that call, and the trumpet will not sound forever. The specific purpose, when it is accomplished, will prove The Presence of God in the world. It comes at the height of one nation's glory and the depth of another people's depravity of condition. He raises one from the valley to the mountaintop; and He lowers one that is on the mountaintop into the valley. This is a picture of the United States of America. America is the greatest and most powerful nation on the Earth. But 400 years ago, White people captured our fathers and brought us to these shores. America is at the height of her power, we are at the depth of a depraved condition, the like of which has never been seen among us as a people. So God's Coming is to America, to take one people down and raise another people up.

The Coming of God is in the person of a human being. This is shocking, because we have always thought God is a spirit. Of course God has spirit—all life has spirit or energy—but we have never thought of God manifesting Himself as one like us, but unlike us. How could He not be what we are, yet make us in His image and after His likeness? We are made in the image and the likeness of God—God is like us and we are like Him. He is our Originator. The difference is that He is Supreme in knowledge, wisdom, understanding and power. The difference is He has the power to make His Word come to pass. He cannot lie. Whatever He desires, He only has to say, "Be!" and it is. The Bible you read now is in English, but it is translated out of the original tongues. What was the original tongue? If the Bible is translated out of the original tongue, and is called the King James Version, who was King James? What kind of human being was King James? Is he qualified to look at the Words of God and give us his version? You have been in enough trouble to know that people can look at the same thing and see it differently. So if there is an accident, and you see it and I see it; and then we are asked what color was the automobile, some might say it was "blue," some might say it was "gray." Some might say it was a sedan; others might say it was a coupe, because emotion, or shock, can damage perception.

My version of the accident may differ from your version of the accident, therefore, there needs to be a judge that listens to the various versions and asks questions to arrive at what are we going to accept as a truthful account of the accident. Here is the "King James Version"; we also have the "Catholic Bible"; the "Scofield Bible"; and we have many different bibles, all of which are translated out of the original tongue. You and I, unfortunately, cannot speak the original tongue, so we have to rely on these translators. The original tongue that the Bible—or Torah and Gospel—was revealed in is Hebrew. Not Greek, but Hebrew. We don't have the original script in Hebrew, so we are relying on King James and his version of what was said originally. Now we have the Holy Qur'an, which was revealed to Prophet Muhammad (Peace Be Upon Him) in Arabic. So, we have an Arabic Qur'an, and this is an Hebrew Torah and Gospel, or Injil. I don't speak Arabic, and I don't speak Hebrew, so I have to rely on a faithful translation. And I must understand that no translation, no matter how faithful, can adequately convey the beauty of the Hebrew language and the Arabic language. I need help in reading and understanding a translation by one who has a knowledge of the original language, one who is wise enough to give me a key, so when I read these books I can make sense out of that which does not easily make sense.

Some of these writers "fixed" the Bible in such a way that the common man would never be able to read it and properly understand the Message that God was trying to convey to us through the mouth of His prophets. I thank Allah (God) for the Holy Qur'an, and for those diligent scholars who studied the English language and tried their best to give a faithful account of Arabic and Hebrew. There's something very interesting about both of these books. These two books are magnificent books, and what I want to do is have a wonderful classroom from the Teachings of the Most Honorable Elijah Muhammad that I want you to reason with. You see, the thing that separates the human being from the vile beasts of the field is our ability to reason. If you feel threatened, reason begins to diminish. If you feel anger, reason begins to diminish. The things that make human beings human and divine is our ability to reason with one another on the basis of truth and perception of truth, so that we can reconcile differences and be of one accord. Now what I want to do today is not to threaten anyone's belief system, but just talk to us; to see if we can reason together on truth. The Bible starts with the first five books of a man called Moses. Although he is a liberator, Moses starts talking about Creation: "In the beginning God created the heaven and the earth. And the earth was without form, and void; and darkness was upon the face of the deep…And God said, 'Let there be light.'"

Who was Moses talking to? Here is a man that came to free some slaves and the first book of the Bible is called Genesis. And in the word Genesis you have the word 'gene,' which has to do with the base, or the beginning, or the birth of something. All of us that are here, had to be born. There was a time when we were confined in a tiny little house called "womb," in darkness. You mothers who have carried your babies to term, you know that time does not start for what is growing in your womb at conception, or your first trimester, or your second trimester, or your third trimester. Once the baby comes out of darkness into light, you call that the birth date. "On such-and-such a day, baby so-and-so was birthed into the world"—that's a genesis. That's a beginning; that's the birth of something. The first book of the Bible is called Genesis, yet Moses was coming to a people who were confined, who were enslaved, who were ignorant, who were in darkness, who had no genesis. Your genesis is not in the time of your mother. Time begins for you when you are freed from your mother. Time begins for a people who are in bondage when they are freed from their oppressor and captor. Since Moses was a liberator, he started with Creation: "In the beginning God created the heaven and the earth. And the earth was without form, and void..." What are you saying, Moses? Just as God created Earth and Heaven out of a void, the Children of Israel had been voided as captives under Pharaoh. You have been voided as a captive under White America.

When you have a check marked "VOID"—I don't care what amount is written on the check; when "VOID" is written across it, it has no value at all. So Moses is saying to the Children of Israel, "As the Earth was without form and void; and darkness was upon the face of the deep, you were without form and void, and darkness was upon the face of your consciousness. But when God said, 'Let there be light'—let there be the introduction of knowledge—He raised me and gave me light to give to a people that walk in darkness." All praise is due to Allah. Now we look at this beautiful book called Bible, which Master Fard Muhammad lovingly called "The Poison Book"—not out of hate, but lovingly. This is a "Poison Book." Well, what is medicine? It's poison—take too much of it, and it will kill you. But if it is prescribed in the right dosage by a competent doctor, and administered by a competent nurse, then the poison becomes medicine that will bring about a good result. However, when you take poison without understanding, something happens to you psychologically, emotionally and mentally. When we open the Bible to this book called Genesis, God is doing all these acts of creation. When it gets to the 26th verse of the first chapter, He's thinking about making man. He said, "Let us make man."

But Moses declared, "Know that the Lord your God is One God." Well, if the Lord your God is One, and God said, "Let us make man," who is the "us" here?

The Holy Qur'an says, "And We made man, and We know what his mind suggests to him." So you have Us in the Bible; We in the Holy Qur'an, yet both the Bible and the Qur'an declare that God is One. Who is Us; and who is We, if God is One? The Semitic languages of Hebrew and Arabic are mathematically precise languages. God is the Revealer of the Word, and He knows these books better than anyone else. He said "We," a pronoun that refers to "groups of persons either in the same room, in the same condition or for the same cause"; and "Us," a pronoun that refers to "a group of people that have one thing in mind"—which is making a man. Neither "We" nor "Us" denies the Oneness of God. Allah (God) is Alone, Singular, Supreme and Above His creation. But when we are in obedience to Him, acting on His Will, then the pronoun "We" is proper because it is still one person acting. It is God, acting through those of us who surrender our will to do His Will, so we become Allah (God). You say, as a Christian, that you are the "body of Christ," but my body cannot function without a head. He is called "Christ" because he has the anointing of the Wisdom, the Power and the Spirit of God. So when my arms act, my feet walk and my tongue speaks, it is not "I" walking; it is not "I" talking—it is God talking through those who surrender to do His Will. All praise is due to Allah. Since Christ is Supreme, except to God; and we are the "body of Christ," then we are supreme beings.

If we are in complete surrender to Allah, we become Allah; we become the supreme beings—the highest manifestation of human behavior. Supreme. God. We. Us. Yet, Allah is One; Indivisible; Independent. This book, Bible, is called scripture. We have Torah; we have Injil, or Gospel; and we have the Holy Qur'an. These are all called scriptures. Even though the

Holy Qur'an is called Al-Kitab (The Book), these two books only deal with 6,000 years of history—but with references to eons of time. The focus of these two books is Adam: The first Adam; the second Adam. A world of evil coming in; a world of evil being destroyed. A new world coming in, and then the Bible ends. In the Holy Qur'an, a prayer opens this book called Al-Fatihah. Wonderful! Seven ayats, or verses, that make up the oft-repeated prayer of the Muslims. The root of that word Fatihah is miftah, which means key. So these seven verses, if understood, give you the key to unlock the knowledge that follows Al-Fatihah. The second chapter of this book is called Al-Baqarah (The Cow). It is in this chapter that Adam is introduced, so both books, in the beginning, start with Adam. Both books tell us that Adam was a powerful being, but he made serious errors. These two books—Bible and Holy Qur'an—deal with a limited time: 6,000 years from the time of Adam to the time of The Coming of God. At the end of the present world, we are entering into the 7,000th year; and the 6,000 years end with a great destruction of the present world.

In the Book of Revelation, you have the 24 Elders; and these Elders are very powerful beings. The Holy Qur'an does not use the term Elders; it uses the term Exalted Assembly. The Honorable Elijah Muhammad taught us—I have to say this because the whole idea of these series of lectures [starting February 25, 2001; through June 3, 2001] is designed to get you to reason with what Elijah Muhammad taught, and to answer the question: "Is it revelation?" If it is revelation, then the Light of God has passed out of the East and is now coming to birth in the West. This means that the scepter of rulership is being taken out of the East and is being placed in the hands of those to whom revelation is given. If a Black man from among us has received revelation from the Lord of the Worlds, this is saying that the White man's time of rule over us and the world is over. Your time is to begin! You have been voided. You are not yet making history. You are inside the womb of America, like a baby is in the womb of its mother, where you are functioning off of his system; feeding from his system, trying to develop a system of your own. But as you begin to evolve, it is not until you are free from him that your genesis will begin. And this is why Moses is given in both Bible and Qur'an as the main man—that the last man would be like "a man like unto Moses." Why? Because you are a people like Moses' people. You and I are a people that walk in darkness. We are without form. We have been made void.

Anything that is voided has no value. You throw it away, you cut it up, you burn it up; or, you put it where you can look at it: A check with a big number on it, that you were never able to cash because it was voided. You are a Black man that doesn't have any value. When the enemy wants to move you, they move you. When they want to kill you, they kill you. When they want to beat you, they beat you. When they want to raise you, they raise you. But it's always when they want it. It is only after you press and press and press, that they give. It's like a baby growing in the womb: Mama may not want to give an inch, but the baby may twist and kick the foot so much that it kicks your nerve, and your legs go out from under you. You didn't mean to lose your balance, but it's that which is in you that is making you lose your balance. So it is with the Black man today. You are moving inside the womb of America—you are not yet free, but your movement is causing trouble inside the womb; inside the nation. The White man is feeling the movement of the Black man, because life is coming into you. And it is only a matter of time before we will all be free.

These excerpts are from Pages 32-42 Part2 of the Origin of Blackness Study Guide 21 The Educational Challenge:
A New Educational Paradigm For the 21 Century
By The Honorable Louis Farrakhan
http://www.finalcall.net/docs/studyguide21.pdf

INDEX

G

Made in United States
Troutdale, OR
05/16/2024

19916842R00056